P9-CML-932

# ASQ Pocket Guide to Root Cause Analysis

# ASQ Pocket Guide to Root Cause Analysis

Bjørn Andersen and
Tom Natland Fagerhaug

ASQ Quality Press
Milwaukee, Wisconsin

American Society for Quality, Quality Press, Milwaukee, WI 53203
© 2014 by ASQ
All rights reserved. Published 2013.


21  20  19  18  17          10  9  8  7  6

Library of Congress Cataloging-in-Publication Data

Andersen, Bjørn.
ASQ pocket guide to root cause analysis / Bjørn Andersen and Tom Natland Fagerhaug.
    pages cm
ISBN 978-0-87389-863-8 (pocket guide: alk. paper)

ISBN 978-1-63694-135-6 (paperback)

1. Total quality management. 2. Problem solving. 3. Quality control. I. Fagerhaug, Tom,
1968- II. Title.
HD62.15.A528 2013
658.4'013—dc23

                        2013034268

ASQ Mission: The American Society for Quality advances individual, organizational, and
community excellence worldwide through learning, quality improvement, and
knowledge exchange.

Attention Bookstores, Wholesalers, Schools, and Corporations: ASQ Quality Press books,
video, audio, and software are available at quantity discounts with bulk purchases for
business, educational, or instructional use. For information, please contact ASQ Quality
Press at 800-248-1946, or write to ASQ Quality Press, P.O. Box 3005, Milwaukee, WI
53201-3005.

To place orders or to request ASQ membership information, call 800-248-1946.
Visit our Web site at www.asq.org/quality-press.

Quality Press
600 N. Plankinton Ave.
Milwaukee, WI 53203-2914
Email: books@asq.org
ASQ  Excellence Through Quality™

# **Contents**

# List of Figures and Tables

# Introduction

Welcome to the pocket guide to root cause analysis! The purpose of this guide is to provide you with easily accessible knowledge about the art of problem solving, with a specific focus on identifying and eliminating root causes of problems. This is a skill that absolutely everybody should master, irrespective of which sector you work in, what educational background you have, and which position in the organization you hold. We hope this pocket guide can contribute to disseminating this skill a little further in the world.

We have previously published two traditional books on the subject of root cause analysis. One, an introduction to RCA, is in its second edition. The other deals with RCA in the healthcare sector specifically. Both were designed to provide practical instruction and advice on how to undertake real-life root cause analyses. It seems logical to take the next step and provide a pocket guide that builds on these books. The strengths of a pocket guide are several: compact presentation of the material, a handy format, and easy access to templates for tools, to name just a few. Readers who are familiar with the original books will find additional value in this pocket guide.

The guide is divided into three main sections:

1. Section one provides a brief introduction to root cause analysis and outlines the RCA process.

2. Section two presents the six steps of the RCA process in detail and describes substeps and available tools and techniques used to accomplish each of these.

3. Section three concludes the guide by giving an example of an RCA project from a manufacturing company.

A pocket guide built on a "proper" book is by definition a condensed version of the original, and our aim for the adaptation has been to preserve a complete overview of the RCA process from start to finish. We often see that potentially successful RCA projects fail when teams charge ahead too quickly and overlook pieces of the puzzle or fail to bring the project to completion by implementing solutions and improvements. We believe this full process view is important.

Another aim has been to make as accessible as possible the various tools and techniques that constitute an important part of RCA skills. You will notice that we provide little preamble or discussion about the tools, but rather give "recipe-like" instructions. If you feel the need to understand more about parts of the RCA process or the approaches employed at the various stages, we suggest our book *Root Cause Analysis: Simplified Tools and Techniques, Second Edition,* ASQ Quality Press, 2006.

**Section I**

## ROOT CAUSE ANALYSIS

All organizations experience unintended variation and its consequences. Such problems exist within a broad range of scope, persistence, and severity across different industries. Some problems cause a minor nuisance, others leads to loss of customers or money, and still others can be a matter of life and death. Anyone will agree that in most cases, preventing problems is preferable to dealing with the consequences of them.

Recurring problems stand out as "sore thumbs" that are most in need of prevention efforts, and root cause analysis can be the key. Examples of problems include:

- A sawmill periodically suffered severe problems of accuracy when cutting lumber to specified dimensions. Experts proposed varying theories as to causes, but the problems persisted. After thoroughly assessing the situation, the parties assigned to pinpoint the reasons for the deviations found the cause to be highly varying air temperature and humidity due to a poorly functioning air conditioning unit.

- Dimensional variation among lamp holders from certain suppliers caused a lot of rework for a lamp manufacturer. Adjustments that needed to be made to ensure proper installation were estimated to cost more than $200,000 annually. Meanwhile, the procurement manager was pleased with himself because he had managed to reduce purchasing costs by about $50,000 the previous year by buying from suppliers that offered the lowest price.

While the terms *root cause* and *root cause analysis* have become part of our business lingo, both carry more meaning than you might expect and both can range broadly in regard to how comprehensively they are perceived. To start with, root cause analysis can be and is practiced as one of two extremes and every shade in between:

- It can be a perfunctory, tedious, form-driven, post–adverse event exercise performed to satisfy some bureaucratic requirement, stealing time and resources that should have been spent doing real work, and not making any difference whatsoever in terms of business results, when we should have just fired the perpetrator of bad practice.

- It can be a motivating, fulfilling, creative exercise initiated because an astute and responsive manager or employee discovered vulnerability in a practice and called together a team to change the process and thereby prevent future negative consequences from recurrence of the problem.

In terms of scope and extent of an RCA project, there can be large variation. A couple of colleagues can easily complete a limited root cause analysis exercise in a few days, changing a faulty practice and solving a problem. The typical project lasts some weeks and involves a small RCA team. Extreme cases can last months or even a year, but these address highly complex problems often requiring investment, organizational change, and training before the root cause is banished. A special type of RCA project is triggered in cases where serious accidents with severe damage to infrastructure, injury, or death have occurred; these often take on the nature of an "investigation" (as in a police-type investigation). Although this latter type is perhaps too rigorous to fit inside the RCA process we outline in this pocket guide, the intention is that the process should work for any type of RCA, from quick and limited to lengthy and comprehensive.

## The Root Cause

Beneath every problem lies a cause. When trying to solve a problem, consider this two-step approach:

1. Identify the cause (or causes) of the problem.

2. Find ways to eliminate these causes and prevent them from recurring.

Depending on the problem, this approach can seem deceptively simple. Indeed, it is easy to underestimate the effort it sometimes takes to find the causes of a problem. Once you've established the true causes, however, eliminating them is often a much easier task. Hence, identifying a problem's cause is paramount. To make things more complicated, a problem is often the result of multiple causes at different levels (see Figure 1). This means that some causes affect other causes that, in turn, create the visible problem. Causes can be classified as one of the following:

- Symptoms. These are not regarded as actual causes, but rather as signs of existing problems.

- First-level causes. Causes that directly lead to a problem.

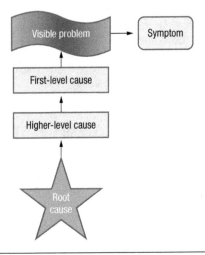

**Figure 1**    Cause levels.

- **Higher-level causes.** Causes that lead to first-level causes. Although they do not directly cause the problem, higher-level causes form links in the chain of cause-and-effect relationships that ultimately create the problem.

Some problems have compound causes, where factors combine. The highest-level cause of a problem is called the root cause; it is "the evil at the bottom" that sets in motion cause-and-effect chains.

## Root Cause Elimination

So how do you ensure that a problem, once it has caused a serious event, does not reoccur next week or next month? Do you simply hope it was a one-off chain of events that will never happen again? The answer is, of course, to remove the root cause. Other approaches might provide some temporary relief, but will never produce a lasting solution.

- If you attack and remove only the symptoms, the situation can become worse. The problem will still be there, but there will no longer be an easily recognized symptom that can be monitored.

- Eliminating first- or higher-level causes can temporarily alleviate the problem, but the root cause will eventually find another way to manifest itself in the form of another problem.

Currently there is no commonly accepted definition of root cause analysis. In general terms, it describes "a process for identifying the basic or causal factors that underlie variation in performance." The meaning of the term *root cause analysis* ranges from a description of a single quality tool to the application of a full problem-solving cycle of improvement. Ideally, root cause analysis is understood as a wide range of approaches, tools, and techniques used to uncover causes of problems and eliminate them.

## THE RCA PROCESS

Conducting a root cause analysis entails a sequence of steps. The step names and substeps vary between users, and we have chosen to describe them based on their most key action. In striving for simplicity, we present the process in this book as a six-step approach. As the subsequent chapters will reveal, there are substeps within each of these, but we believe that limiting the main steps to six will make understanding and remembering the process easier.

### Root Cause Analysis Steps

Briefly, a typical RCA follows a series of six steps (Figure 2):

1. Define the event, succinctly describing the event or deviation that triggered the RCA.

2. Find causes, coming up with as broad a range of potential causes of the problem as possible.

3. Find the root cause, zooming in on the main culprit.

4. Find solutions to solve the problem and prevent the event from reoccurring.

5. Take action, implementing solutions to ensure that things stay that way.

6. Measure and assess to determine whether the solution(s) work and solved the problem.

**Figure 2** Root cause analysis steps.

Table 1 provides more information about the six steps, detailed under these headings:

- Purpose. This column provides key words to describe why the step is included in the RCA process. For some steps, there are probably additional purposes, but we have tried to limit lists to the most important.

- Output. Lists the main "products" that should result from the execution of a step, again limited to the key elements.

- Characteristic. This is an attempt to indicate whether each step is typically analytical or creative in nature. It is perhaps a bit ambitious to label each step one or the other; some steps will exhibit both characteristics. However, this identifies the main mode of thinking for a step.

- Percent of duration. This is probably the most imprecise of all the columns. It is virtually impossible to provide general and accurate estimations of how long a step in the RCA process will take; duration will vary dramatically depending on the type of event or problem being addressed. These are estimates for a "typical" RCA project.

- Success factors. This column includes key factors that should be observed so as to successfully complete a step.

- Tools. The basic tools presented in this guide can be used at each stage of the RCA process. More comprehensive books on RCA will include additional tools.

| Step | Purpose | Output | Characteristic | % of Duration | Success Factors | Tools |
|------|---------|--------|----------------|---------------|-----------------|-------|
| 1. Define the event | • Scope the problem<br>• Provide starting point<br>• Organize RCA team<br>• Create realistic project plan | • Problem definition<br>• RCA team<br>• Project plan | Analytical | 5–10 | • Being specific and objective<br>• No speculation about what caused the event<br>• Relevant team composition<br>• Schedule balances ambition and time | • Gantt chart<br>• Interview<br>• Survey |
| 2. Find causes | • Understand the problem better<br>• Create a broad overview of possible causes<br>• Ensure that all involved are heard | • List of possible causes | Creative | 10–15 | • Access to background data and evidence about the event and problem<br>• Being able to think creatively<br>• No sorting/ screening of suggestions for possible causes | • Flowchart<br>• High-level mapping<br>• Brainstorming<br>• Fishbone diagram |

**Table 1**   Root cause analysis steps.

*(Continued)*

| Step | Purpose | Output | Characteristic | % of Duration | Success Factors | Tools |
|------|---------|--------|----------------|---------------|-----------------|-------|
| 3. Find the root cause | • Uncover the true root cause leading to the event/problem | • Description of the root cause | Analytical | 20–30 | • Keeping a calm and analytical mind<br>• Do not declare finding the root cause too soon<br>• Dare to call things by their real name | • Cause-and-event tree<br>• Five whys<br>• Fault tree<br>• Pareto analysis<br>• Scatter chart<br>• Histogram<br>• Problem concentration diagram |
| 4. Find solution(s) | • Design workable solution(s) that eliminates the root cause | • Description of the solution(s) | Creative | 15–25 | • Involving those with ideas about possible solutions<br>• Involving those who will be affected by the solution<br>• Creative ownership of the required changes | • Flowchart<br>• Interview<br>• Survey<br>• Brainstorming<br>• Benchmarking<br>• The "Why Not" principles |

*(Continued)*

**Table 1**   Root cause analysis steps.

| Step | Purpose | Output | Characteristic | % of Duration | Success Factors | Tools |
|------|---------|--------|----------------|---------------|-----------------|-------|
| 5. Take action | • Implement the solution(s)<br>• Ensure lasting changes in practice | • Implemented solution(s) | Analytical | 5–50 | • Involving those who will have to change their work processes<br>• Be patient and persistent | • Impact effort matrix<br>• Force field analysis |
| 6. Measure and assess | • Assess the effectiveness of the implemented solution(s)<br>• Review whether further effort is required<br>• Close the RCA project | • Solution(s) confirmed to solve problem<br>• Project report | Analytical | 5–10 | • Be critical in the assessment of the solution(s)<br>• Don't be too eager in declaring success and closing the project | • Pilot study |

**Table 1**  Root cause analysis steps.

We should also point out that although this is presented as a linear and sequential process, in reality there will often be cases of both overlapping steps and loops where it is necessary to repeat earlier steps and do more work before proceeding.

## The Logistics of a Root Cause Analysis

As an example, the following represents a typical organization of a root cause analysis project:

- A small team is formed to conduct the root cause analysis.

- Team members are selected from the business process/area of the organization that experiences the problem, supplemented by a line manager with decision authority to implement solutions, an internal customer from the process with problems, and possibly a quality improvement expert in the case where the other team members have little experience with this kind of work.

- The analysis lasts about two months, with time relatively evenly distributed between defining and understanding the problem, brainstorming possible causes, analyzing causes and effects, and devising a solution.

- During this period, the team meets at least weekly, sometimes two or three times a week. Meetings are kept short, a maximum of two hours, and because meetings are meant to be creative in nature, the agenda is quite loose.

- One person is assigned the role of making sure the analysis progresses; other tasks are assigned to various members of the team.

- Once the solution has been designed and the decision to implement has been taken, it can take anywhere from a day to several months before the change is complete, depending on what is involved in the implementation process.

**Section II**

### STEP 1: DEFINE THE EVENT

The first of the six steps in root cause analysis is to define the event, succinctly describing the deviation that triggered the RCA. In our experience, it is important to address this step properly because it is vital that the team have a common understanding of the event before pursuing the rest of the RCA process.

### Purpose of Step 1: Define the Event

- The overall purpose of this step is to provide an unambiguous starting point for the root cause analysis process by scoping the problem.

- The step includes organizing the RCA team and creating a realistic project plan.

- The outputs of this phase are a problem definition, an RCA team, and a project plan.

**Success Factors for This Step**

- Being specific and objective; that is, call things by their real names and don't be afraid to discuss sensitive issues

- No speculation about what caused the event—this comes later

- Relevant team composition, ensuring access to required knowledge and ownership of the process and solutions

- A schedule that balances ambitions and time, allowing for sufficient debate

- Not placing any blame or speculating "whose fault it was"

## Substeps in Step 1: Define the Event

a. Trigger the RCA process, that is, officially launch the RCA project.

b. Mandate and organize the RCA team. This can seem somewhat bureaucratic, but the mandate is important in providing the team with the necessary authority to collect evidence and data, propose solutions, and act. Organizing also involves resourcing, that is, appointing team members and ensuring their availability.

c. Plan the RCA project by defining the tasks to be performed, by whom, by when, and defining milestones.

d. Describe the event in detail; this is probably the most demanding of these substeps.

## 1a: Trigger the RCA Process

To some extent, triggering the RCA process is not a step per se; an event happens and creates the need for the analysis. On the other hand, someone in the organization must step up and formally launch it. Who that "someone" is depends on the event; it could be a unit manager, an administrator, a quality manager, and so on.

*Trigger sources*

A typical RCA process could be triggered by a variety of sources:

- Internal triggers: employees observing poor practices, or someone having witnessed an event

- External triggers: customers, suppliers, media, or other stakeholders

- System triggers: reviews, surveys, or audits

- Specific incidents: employee/customer injuries or fatalities, damage to equipment, or other events that exceed a predetermined limit

*Trigger examples*

A major retailer and manufacturer of home furnishing products learns that a certain glass shower enclosure sometimes breaks. Several factors might lead to initiation of an RCA process:

- An employee at the service desk meets a customer whose shower enclosure has broken, and where the smallest child had several glass injuries.

- An employee working with return statistics sees a number of similar occurrences worldwide and believes that the probability of related injuries is significant.

Based on this input, the head of the customer service department launches an RCA process.

## 1b: Mandate and Organize the RCA Team

Based on these substeps, a clear and concise mandate should be developed in concert with organizational policy. The mandate should define the team's authority, responsibility, and objectives, the latter typically being to identify the root cause and recommend how to eliminate it. Figure 3 illustrates a team mandate template.

### *RCA Team Composition*

A typical RCA team should be made up of the following participants:

- A team leader who has substantial knowledge about the event and authority in the organization

- A facilitator, who should be experienced in conducting RCA and facilitating teams

- Team members, normally a maximum of six

See Table 2 for team composition template.

# RCA Team Mandate

Date: _____ Signature: _____

Event to investigate: _____

Objectives for the RCA:
1. _____
2. _____
3. _____

Team authority:
1. _____
2. _____
3. _____

Team responsibilities:
1. _____
2. _____
3. _____

Special conditions: _____
_____

**Figure 3**   RCA team mandate template.

| Team member no. | Role | Name | Organizational unit | Position | E-mail | Phone |
|---|---|---|---|---|---|---|
| 1 | Leader | | | | | |
| 2 | Facilitator | | | | | |
| 3 | Member | | | | | |
| 4 | | | | | | |
| 5 | | | | | | |
| 6 | | | | | | |
| 7 | | | | | | |
| 8 | | | | | | |
| 9 | | | | | | |
| 10 | | | | | | |

**Table 2**   RCA team composition template.

*RCA Team Members Criteria*

RCA team members should share the following characteristics:

- Time to participate actively and wholeheartedly in the work.

- Knowledge about the organization and the process where the event occurred, and training in root cause analysis.

- Motivation, that is, a desire to eliminate the problem and create improvements. A person selected against his or her will is an unsuitable member of an RCA team.

- The ability to cooperate, listen, and communicate; root cause analysis is typically a team effort that is not suited for an introverted participant.

- Credibility and respect in the organization, to ensure impact when presenting results from the project and proceeding with effective implementation of improvements.

## 1c: Plan the RCA Project

Once the mandate and team composition are set, it is time to plan the project. The plan must reflect both internal ambitions and external requirements (if applicable). Based on these expectations and requirements, a detailed plan should be developed defining tasks, responsibilities, resources, sequence, and milestones. The project plan is usually visualized by means of a Gantt chart. To the extent possible, we recommend that the plan be based on the steps and substeps in this book.

*Gantt charts*

Probably 95 percent of all project plans are made using Gantt charts—a good indication that it is a useful tool. The main purpose of a Gantt chart is to depict the project tasks and the schedule and to provide the basis for monitoring progress. The steps in creating a project plan with a Gantt chart (using paper, spreadsheet, or dedicated software) are shown here:

1. List all tasks or activities.

2. For each activity, define the latest finish date and earliest starting date, the duration, and any dependence on other activities.

3. Place the tasks in an empty Gantt chart, with the timeline reflecting the overall duration of the project.

4. Schedule activities by drawing bars that correspond to the duration of the activities.

5. Use diamonds to depict milestones in the plan.

*Gantt chart example*

A manufacturing firm experienced a significant rise in employee sick leave requests. A project team was assembled, and a plan was outlined using a Gantt chart (Table 3). Table 4 is a Gantt chart template.

| Activity # | Activity | Responsible | Duration | W8 | 9 | 10 | 11 | 12 | 13 | 14 | 15 | 16 | 17 | |
|---|---|---|---|---|---|---|---|---|---|---|---|---|---|---|
| | Define the event | Debbie | 1 week | ▓ | | | | | | | | | | |
| | Find causes | John | 2 weeks | | ▓ | ▓ | | | | | | | | |
| | Find the root cause | Ray | 3 weeks | | | | ▓ | ▓ | ▓ | | | | | |
| | Find solution(s) | Tina | 2 weeks | | | | | | | ▓ | ▓ | | | |
| | Take actions | Tina | 2 weeks | | | | | | | | | ▓ | ▓ | |
| | Measure and assess | Debbie | 1 week | | | | | | | | | | | ▓ |

**Table 3**  Gantt chart example.

| Activity # | Activity | Responsible | Duration | | | | | | | | | |
|---|---|---|---|---|---|---|---|---|---|---|---|---|
| | | | | | | | | | | | | |
| | | | | | | | | | | | | |
| | | | | | | | | | | | | |
| | | | | | | | | | | | | |
| | | | | | | | | | | | | |
| | | | | | | | | | | | | |
| | | | | | | | | | | | | |
| | | | | | | | | | | | | |
| | | | | | | | | | | | | |
| | | | | | | | | | | | | |

**Table 4**  Gantt chart template.

## 1d: Describe the Event in Detail

Probably the most challenging substep is to arrive at a detailed, precise, and unambiguous description of the event. The description should at a minimum address the following:

- What is the event?
- When did it happen?
- Where did it happen?
- Who was involved?
- Has it happened before? If so, how often?
- What were the consequences of the event?

Some advice about this task: provide specific details instead of symptoms, eliminate bias stemming from personal emotions or involvement in the event, and keep suspected causes out of the event statement.

*Find the core*

The aim of this exercise is to move from a "big hairy problem" to a precise and objective description of the event, as illustrated in Figure 4.

Consider the difference between these two statements regarding window repair shop appointments that take longer than planned:

- Solution: Set aside more time for each customer (car).
- Event: Customer not notified in advance regarding the different options regarding new front window in car.

"Hairy problem"

Precise description of event

**Figure 4**   Precise description of event.

Both describe a situation where changing a front window takes too long. Depending on how the problem is described, the remedy takes on quite a different nature.

*Eliminate bias*

A precise description of an event depends on the elimination of bias and emotion. It's important to avoid thinking about causes at this stage. Both the RCA team and stakeholders will harbor emotions and views about what happened, especially those affected by the event. The RCA should strive to eliminate bias by focusing on the problem and continuously striving for objectivity.

*Collect data*

In moving from a "big hairy problem" to a precise description, you can use one of several approaches to collect information and data. We have chosen two that we consider the most universal for this purpose: interview and survey.

Interview. When you want to gather verbal information, interviews are a powerful approach. The main purpose of interviews is to gather information from those involved in the event, either directly or indirectly. Interviews may be used at different stages of the RCA process. The main steps in using interviews are these:

1. Prepare an interview guide.

2. Test the questions in order to eliminate ambiguity.

3. Make an appointment with the people you want to interview.

4. Make sure you have privacy and are not disturbed during the interview.

5. Ask the questions and make sure they are understood.

6. Record the answers, digitally or in writing.

7. If relevant, obtain the interviewee's confirmation that the answers were understood correctly.

Survey. When you want to collect data about people's attitudes, feelings, or opinions, surveys are a useful approach. The main purpose of surveys is to collect data from a large number of respondents. In root cause analysis, the most common use of surveys includes collecting patient satisfaction data and employee attitude data from units where an event has occurred. The steps in carrying out a survey are typically these:

1. Define the objective of the survey and what information is required to achieve this objective.

2. Decide how the survey will be undertaken: written (via mail, e-mail, or online) or verbal (by telephone or in person).

3. Develop the questionnaire and test it with employees uninvolved in the survey.

4. Identify the sample of respondents and send the survey to them.

5. Collect the data according to the chosen approach and analyze it.

Survey Example. A computer store had specialized in selling to unskilled buyers, some of whom were touching a computer for the first time. Many customers required a lot of support and technical guidance during the first few weeks after purchase, and many complained about their buying experience. To determine what caused these problems, the store developed a customer satisfaction survey; they sent a simple questionnaire (Figure 5) to every buyer six weeks after the purchase, along with a postage-paid return envelope. To encourage people to return the questionnaire, the store added respondents' names to a drawing for $1,000 in software. The survey yielded about 150 completed questionnaires. After company officials assembled and analyzed the data, the cause of most dissatisfaction became clear.

# Customer Satisfaction Survey

To improve our service to you, we are conducting a small survey on your experience in buying a computer from us. We would highly appreciate your taking time to fill in this questionnaire.

Please indicate your responses by checking the appropriate boxes.

|  | Poor | | | | | Excellent |
|---|---|---|---|---|---|---|
|  | 1 | 2 | 3 | 4 | 5 | 6 |

1. Overall, how would you rate your purchase from our store? ............... ☐☐☐☐☐☐

2. How would you rate the following aspects of our service?

   Computer hardware and accessories selection............................ ☐☐☐☐☐☐

   Hardware and accessories prices.......... ☐☐☐☐☐☐

   Software selection ...................... ☐☐☐☐☐☐

   Salesperson's knowledge and ability to help you ........................... ☐☐☐☐☐☐

   Delivery time of the equipment you bought.. ☐☐☐☐☐☐

   Quality of the instructions and manuals..... ☐☐☐☐☐☐

   Technical support during installation ....... ☐☐☐☐☐☐

   Technical support after first installation ..... ☐☐☐☐☐☐

   Reliability of the equipment............... ☐☐☐☐☐☐

3. Would you recommend our store to others?  ☐ Yes  ☐ No

4. What is your age?  ☐ <30  ☐ >30

5. What is your gender?  ☐ Male  ☐ Female

**THANK YOU VERY MUCH!**

**Figure 5**  Customer satisfaction survey.

## Checklist for Step 1: Define the Event

- ☐ The RCA process has been officially triggered.
- ☐ An unequivocal mandate for the RCA process and team has been produced.
- ☐ An RCA team with sufficient competence and resources has been appointed.
- ☐ A detailed plan including responsibilities, resources, sequence, and milestones has been developed.
- ☐ A detailed, precise, and unambiguous description of the event has been made.
- ☐ Where relevant, required data about the event have been collected.

## STEP 2: FIND CAUSES

This section presents the detailed steps of the second phase of the RCA process, which revolves around generating an overview of the possible causes leading to the event under investigation.

## Purpose of Step 2: Find Causes

- The main purpose of Step 2 is to generate as extensive a list as possible of potential causes that could have led to or contributed to the occurrence of the event. This will ensure a good starting point for the exercise of identifying the root cause.

It's important to ensure that this list of possible causes covers ideas and input from all stakeholders who might have knowledge about the event and its causes. This is critical for at least two reasons: you avoid overlooking possibly important causes and you counter any tendency toward resistance to future solutions.

**Success Factors for This Step**

- Having access to sufficient amounts of background data and evidence about the event or problem.

- Generating as broad a set of potential causes as possible.

- Allowing free and creative thinking. Curbing people's enthusiasm and "crazy thinking" will cause this step to fail.

- No sorting/screening of suggestions is allowed. This is a golden rule of brainstorming and it certainly applies here; if you allow criticism or elimination of ideas during the creative phase, people will naturally hold back and ideas will be missed.

## Substeps in Step 2: Find Causes

a. Map the sequence of activities (process) within which the event took place

b. High-level mapping of the context of the event and the process it occurred within

c. Brainstorm a wide range of possible causes of the event, either through regular brainstorming or aided by a fishbone diagram

## 2a. Map the Event

The mapping process exists inside an organization that may be composed of several entities; it has stakeholders and it operates in an environment conditioned by different contextual factors (economics, incentives, regulations, and so on). There are really two levels of mapping involved (Figure 6); first the event itself is depicted as an encapsulated object and secondly, factors surrounding the process being performed when the event occurred:

- The boundaries around the event are defined in Step 1 of the RCA process.

- The event took place in a sequence of activities (often termed a process).

- The steps leading up to the event and being carried out after the event must be understood.

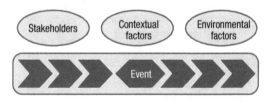

**Figure 6**    Mapping levels.

*Flowcharting*

Such process mapping is almost always carried out using flowcharts and following a few simple steps:

1. Gather those employees working in the process in a meeting room with whiteboard facilities and plenty of adhesive notes in different colors.

2. Define the start and end points of the process as well as boundaries between parallel processes.

3. Identify the main activities or tasks undertaken during the process (sometimes it is useful to start with the final outcome and work backward).

4. Create adhesive notes in different colors to represent activities, products, documents, and other elements of the process.

5. Map the process by moving the notes around on the board until they reflect the most realistic picture of the process in question.

6. Butcher paper is helpful in case the flowchart of the process becomes too large to be easily readable on a standard page.

*Flowchart symbols*

People use a range of different symbols when constructing flowcharts. Some are more universally agreed upon than others, and some are standardized in various software packages for flowcharting. We encourage you to agree on a set of symbols understood in your organization, and we provide some examples of common symbols in Figure 7.

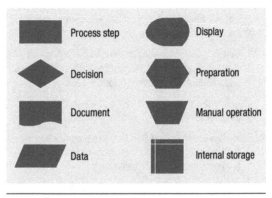

**Figure 7**　Flowchart symbols.

*Flow chart example*

A small engineering firm, Leaner & Smarter, had grown significantly the last year and thus had to better describe its management system and underlying processes. Additionally, new customers were requiring that Leaner & Smarter obtain an ISO 9000 certificate within the next half year in order to be allowed to submit tenders. Leaner & Smarter analyzed and documented their procedures, comparing these with the ISO requirements, and found that their procedure for discipline control (DIC) had to be updated. They gathered their engineers and came up with the basic procedure shown in Figure 8.

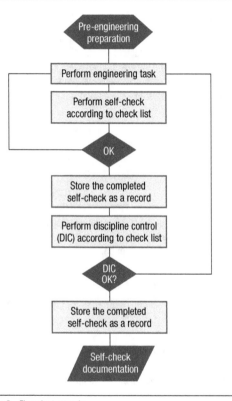

**Figure 8** Flow chart example.

## 2b: High-Level Mapping

Having mapped the process containing the event, you will often realize that other factors surrounding the process affect both the process and the event under investigation. These can be uncovered through a higher-level mapping, the purpose of which is to understand issues and forces influencing the process:

- Stakeholders. External or internal individuals or organizations with a vested interest in the process (for example, patients, regulatory bodies, families, service or equipment suppliers, insurance companies, and employees)

- Contextual factors. Factors that define framework conditions for the process (for example, financial situation, availability of resources, and incentives driving certain types of behavior)

- Environmental factors. Factors that describe the environment in which the process is being performed (for example, temperature, level of sterility, capacity utilization, and stress)

*Contextual and environmental factors*

Contextual and environmental factors might seem quite abstract and difficult to relate to. They could be, but the purpose of this exercise is simply to put words to issues that somehow shape and influence the process and setting where the event took place.

*Contextual and environmental factor examples*

Having seen an increasing number of cases where patients' use of herbal supplements caused surgical complications such as bleeding, cardiac arrhythmia, and other complications, a hospital team discussed other factors that influenced these cases:

- Lack of resources, especially nurses, to perform perioperative interviews.

- Nurses' preference to be involved in surgical procedures rather than perioperative preparations.

- Perioperative interviews are often conducted as telephone interviews, but the interviewers have few quiet places suitable for undertaking the interviews, thus motivating them to make them as short as possible.

- Language barriers between interviewers and patients lacking English skills, which made it difficult to pose questions and understand responses.

- Lack of industry and organizational knowledge regarding the complications associated with non–FDA approved supplements.

- Reluctance of patients to disclose the quantity and kind of nonprescription supplements that they use.

## 2c: Brainstorm Possible Causes

Brainstorming is quite possibly the most widely used "tool" in organizations around the world. As such, it is probably well known to many readers. The purpose of brainstorming is simply to come up with as many ideas as possible, including "crazy ideas," about possible causes for the event being analyzed. The steps are:

1. Acquire a whiteboard or flip chart to record ideas.

2. Open the floor to participants for launching ideas, encouraging everyone to participate.

3. Write down every idea launched, using the same wording as the original proposition.

4. Do not discuss, criticize, or evaluate ideas during the session.

5. Allow the flow of ideas to stagnate once because it will usually pick up again; close the process when few new ideas emerge.

6. Evaluate ideas by sorting them into groups of decreasing relevance.

*Fishbone diagram*

The fishbone diagram is a tool used to understand relationships between a problem or event and its causes (Figure 10). It is a technique that aids brainstorming. The approach follows these steps:

1. Using a whiteboard or similar surface, place the event at the right end of a large arrow.

2. Identify main categories of causes and write them on lines branching off from the large arrow.

3. Proceed through the chart, one main category at a time, and brainstorm all possible causes, placing them on the relevant branches.

4. Use brief and succinct descriptions of causes. Write causes that belong to more than one category on all relevant branches.

*Fishbone diagram example*

A company operating cable television services had seen consistently high employee absenteeism, especially in the installation and service department. Besides costing the company a lot of money, this absenteeism angered customers because hook-ups were not done at the agreed time and problems took an unacceptably long time to correct.

A fishbone diagram was constructed containing many ideas as to why absenteeism was so high (Figure 9). The results led the company to consider training programs, reward systems, and the quality of tools and equipment used by the service personnel.

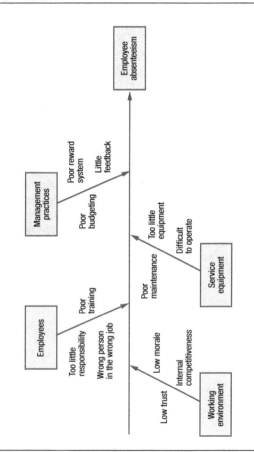

**Figure 9**    Fishbone diagram example.

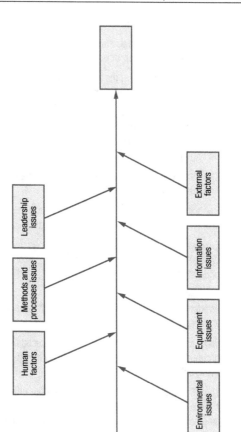

**Figure 10** Fishbone diagram template.

## Checklist for Step 2: Find Causes

- ☐ The effort to find causes started only after the event or problem had been precisely defined.

- ☐ The process containing the step or activity where the event occurred has been mapped.

- ☐ Participants in the process mapping included people involved in performing the process, and also people delivering input to it, receiving output from it, and managing it.

- ☐ The stakeholders of the process have been mapped, and their expectations analyzed.

- ☐ Contextual factors defining the setting where the process is being performed have been understood.

- ☐ Environmental factors defining the conditions under which the process is being performed have been understood.

- ☐ Possible causes for the event have been brainstormed, without any criticism or discussion of ideas as they were launched.

- ☐ Ideas for possible causes have been discussed and grouped according to assumed relevance for the event being analyzed.

- ☐ The RCA team is content that all relevant possible causes have been identified and is ready to leave this stage of the process to proceed with looking for the root cause.

## STEP 3: FIND THE ROOT CAUSE

This section presents the detailed steps of the third phase of the RCA process, which focuses on finally identifying the actual root cause.

### Purpose of Step 3: Find the Root Cause

- Being able to stay with this change of thinking, from creative to analytical

- Staying the course and resisting declaring success too early

- Ensuring an open climate in the RCA team, where causes and connections are openly discussed and called what they are

### Success Factors in This Step

- Creativity versus analysis. If you succeeded in the previous step, having found a broad set of possible causes, it is now more fruitful to focus on analyzing these and winnowing them down to the actual root cause rather than reopening the discussion for further brainstorming.

- Not declaring success too soon. In some cases you believe the root cause has been found, when in fact there are further levels of cause below it. Sometimes it can be quite difficult to know where to draw the line; we often find that RCA teams stop too soon.

- Drill down beyond individual blame and address the conditions that allowed the event to occur. There will be situations where intermediate or even root causes come down to human error, forgetfulness, lack of skills, and so on. Especially in such cases it is critical to find the root cause in the system that does not support human frailties and creates conditions ripe for human error.

## Substeps in Step 3: Find the Root Cause

a. Categorize and group possible causes from step 2

b. Construct a cause-and-event tree

c. Analyze possible causes to identify the root cause

d. Collate the findings and revise the cause-and-event tree

Some events are easier to analyze than others; if the cause-and-effect tree in step 3b already reveals the root cause, there is no need to run through all the substeps.

### 3a: Categorize and Group Possible Causes

In most RCAs, if not all, the list of possible causes contains a wide variety including technical issues, procedural glitches, human factors, systemic elements such as money or incentives, and so on. Analysis is much easier if you bring some order to the list first. This is done through two operations:

- Categorize the possible causes in logical subsets (possible subsets are outlined at right).

- Group possible causes that seem to be similar, overlapping, or in other ways related.

Please notice that the categorization effort often leads to the identification of some new possible causes. This is quite natural and these should be included in the analysis.

*Possible cause categories*

There are many ways to categorize possible causes. We have found the following generic list to be useful:

- Environmental issues. Causal factors found in the environment where the event took place such as temperature, noise, clutter.

- Equipment issues. Errors or problems with all types of equipment being used.

- Methods and processes. Issues pertaining to the different processes and procedures being run in the organization.

- Human factors. All issues related to human effort or intervention in a process.

- Leadership issues. Causes resulting from the climate and culture created by the organization's management.

- Information issues. Causes linked to lack of information or erroneous information.

- External factors beyond control. Any causes that are beyond the control of the organization, for example, weather, regulations, supplier strike.

## 3b: Construct a Cause-and-Event Tree

A cause-and-event tree is used to analyze various ways problems and events can occur in a system. We use some of the same approach here, mostly to create an overview of where the analysis stands at this stage. The purpose of this exercise is to:

- Undertake a first pruning of the possible causes

- Create a first insight into the hierarchical connections between the identified possible causes

- Allow a visual portrayal of the categories and groups of possible causes found

*Cause-and-event tree example*

After a hospital patient developed a pressure ulcer, an RCA team was mandated to look into the event. The team came up with a number of possible causes and realized that these were partly related; they created a cause-and-effect tree (Figure 11) to understand linkages among the possible causes. The tree showed multiple linkages and helped point the RCA team in the direction of likely root causes.

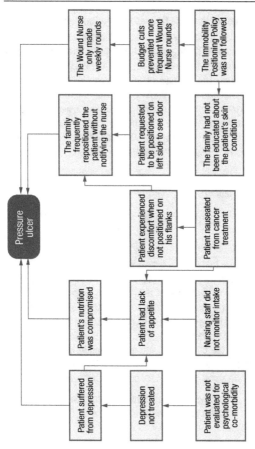

**Figure 11** Cause-and-effect tree.

## 3c: Analyze Possible Causes to Find the Root Cause

So far, we have presented the RCA process as quite linear and stepwise. When you reach this point, the road ahead is determined by the event in question and the possible causes uncovered. The purpose of this substep is to delve deeper into the possible causes and zoom in on the root cause. This is normally done through the use of various analysis tools and techniques. Sometimes one analysis is enough; often several tools must be applied. The selection of useful tools is extensive; here we will present some of them:

- Five whys
- Fault trees
- Pareto analysis

- Scatter charts
- Histograms
- Problem concentration diagrams

*Choosing Your Tool*

We realize that providing several alternative tools for this task can be confusing! First of all, let us assure you that very often you will find that none of these are necessary to use. If you find that the possible causes you have come up with are clear and point in the direction of a root cause, consider yourselves lucky and move on.

If, on the other hand, you see a need to look further into possible causes, one of these techniques could be helpful. Here are some guidelines for choosing between them:

- *Five whys* is the most fundamental of all root cause analysis tools. If you use nothing else, use this technique.

- If you have identified a large number of possible causes, and especially if these seem to belong to different "branches" of cause-and-effect chains, a *fault tree* can be quite useful in visualizing the branches.

- If there is a geographical dimension to the event and its causes (that is, occurrences of the event are scattered over an area, or causes are found here and there), a *problem concentration diagram* can help identify patterns or concentrations of issues.

- If some of the possible causes you identify are beyond what the RCA team or even the institution can do something about, a *span of control analysis* can sort out which causes to attack for maximum benefit.

### Five Whys technique

The Five Whys technique, also known as the why–why chart, is the quality management field's equivalent of a dentist's drill; its inherent nature is to penetrate deeper and deeper into "where it hurts," that is, the underlying root causes. Its main purpose is to constantly ask "Why?" to make sure that you don't stop before uncovering the true root cause. It uses the following steps:

1. Determine the starting point—here an assumed root cause—and write it at the top of a flip chart or whiteboard.

2. Ask, "Why did the root cause occur?" If an answer surfaces that is in itself a cause leading to the assumed root cause, the true root cause has not been found, but rather just a proximate cause.

3. Put the new cause below the originally assumed root cause.

4. Repeat the "why" question, continuing until no new answer results. The last answer will most likely be the actual root cause. (Starting from the original event, as opposed to an assumed root cause, this method often requires five rounds of why until the chain reaches the end; thus the name of the technique, although five is certainly not an upper limit.)

*Five Whys example*

As a small business in the rapidly growing world of web site design and programming, an enterprise of about 25 people had grown from a small home-based outfit into the current company with many large business clients.

Previously, the team of web programmers had received much acclaim for web page design and innovative use of graphics to make sites easy to navigate. Lately, however, more clients were dissatisfied with the web sites. They complained about functionality, simple errors in layout or text, late completion of designs and entire sites, and so on.

The situation had gotten to a point where the employees faced constant problems and no longer thought of the work as fun. Some of the entrenched technology freaks blamed the company's unwillingness to stay abreast in this development; others thought most of the problems stemmed from the lack of qualified programmers.

To get to the bottom of this problem, which started to threaten the future of the company, one of the founding partners used the Five Whys tool. The resulting chart and a template are shown in Figure 12. As you can see, the root causes were neither of those previously believed to be the culprits, but rather too many projects being undertaken simultaneously.

| Dissatisfied web site customers |

Why?  Lacking functionality

  Why?  Poor customer communication

    Why?  Too much time pressure

      Why?  Too many projects

| Template |

Why?

  Why?

    Why?

      Why?

        Why?

**Figure 12**  Five Whys example and template.

*Fault tree analysis*

A fault tree is used to progress beyond the cause-and-event tree by being more specific about the connections between causes and the event. The steps in building a fault tree are:

1. Place the event at the top of a tree diagram. Sometimes the diagram is constructed "vertically" and thus shaped like a Christmas tree, wide at the bottom and tapering to the top. Sometimes it is rotated 90 degrees for space purposes and expanding to the right (as in Figure 13).

2. Put immediate causes at the level just below the event.

3. For each cause, assess whether it is the result of lower-level causes or represents a basic cause. Draw circles around basic causes not to be developed further; draw rectangles around intermediate causes.

4. Continue until the diagram contains only basic causes at the lowest level of each branch.

5. Where more than one cause leads to the level above, use symbols to connect the branches to indicate whether these operate together (AND, symbolized by ◓ ) or on their own (OR, symbolized by ▲)

*Fault tree analysis example*

When the web design company of the Five Whys example realized that many of their problems were caused by taking on too many assignments at the same time, a whole new area of causes was opened up. The five whys analysis was followed up using fault tree analysis, both to generate further causes to the problems and relate these to each other. The resulting fault tree (Figure 13) helped the company understand that several different causes contributed to the problem:

- Too many projects taken on

- Lack of a good project management system

- Poor organization of the resources and their work

*Pareto analysis*

The Pareto principle states that most effects, often 80 percent, are the result of a small number of causes, often only 20 percent. Pareto analysis tries to identify these few causes, as these are likely candidates for root causes. The analysis can be carried out by using either a list of causes or a table where the causes are sorted, or by placing the causes in a chart:

1. Start with the potential causes that have been identified.

2. Decide which criteria to use when comparing the possible causes (for example, how often they occur, their consequences, or costs).

3. If data required for chosen criteria do not already exist, collect them.

4. Sort the causes according to descending scores for the criteria.

5. Present the causes in this order in a list, table, or chart, showing absolute and cumulative data for each cause. The cause with the highest score, say 40% of the consequences, is listed first; the next, representing for example 25%, brings the cumulative total to 65%, and so on.

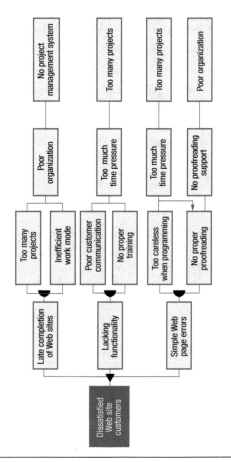

**Figure 13**　Fault tree analysis example.

*Pareto analysis example*

Many studios around the world make television commercials. One studio specialized in shooting ads starring cats. This proved very popular and the company prospered. Lately, though, many shoots were taking much longer than planned, causing production delays despite the use of overtime and weekend work. These delays were related to lack of equipment, technical problems with audio and video, rework of the script, and misbehaving cats.

In fact, misbehaving cats seemed to be the dominant problem area, and it was decided to identify what seemed to cause the unrest. The set assistant on duty was to record what he or she believed to be the reason the cats caused problems and filled quite a few pages with notes. Some of the data are shown in Table 5. Not knowing exactly how to attack this data, someone recommended using a Pareto chart to determine the prevailing causes. The analysis led to changes related to the scheduling of shootings and the preparation of the cats (Figure 14). The template for a Pareto chart is shown in Figure 15.

| Cause of cat distress | Time lost due to the cause (minutes) | Total time lost due to the cause (minutes) |
|---|---|---|
| Not been fed | 4, 3, 5, 2, 5, 3 | 22 |
| Not been cuddled | 3, 3, 5, 3 | 14 |
| Studio too cold | 9, 2, 4, 6, 4, 5 | 30 |
| Too much noise | 20, 15, 35, 20, 9, 16 | 115 |
| Smell of previous cat still present | 41, 68, 39, 60, 29, 52, 19, 8 | 316 |
| Surface to sit/lie on not appealing | 2, 4, 1 | 7 |

**Table 5**   Cat studio data.

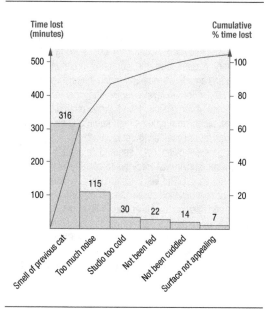

**Figure 14**   Pareto chart example.

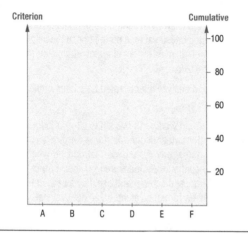

**Figure 15** Pareto chart template.

## *Scatter charts*

Causes at different levels often impact one another. A scatter chart (Figure 17 on page 60) can identify such links. A prerequisite is that each cause can be expressed by a numerical value. The main purpose of the scatter chart is to show the relationship between two causes or other variables. The steps in creating a scatter chart are as follows:

1. Select the two variables to be examined (one dependent and one independent).

2. For each value of the independent variable, measure the corresponding value of the dependent variable.

3. Plot the values from each data pair on the chart.

4. Draw the chart by placing the independent, or expected cause variable on the horizontal axis, and the dependent or expected effect variable on the vertical axis.

5. Plot and analyze the collected data pairs in the chart.

*Scatter chart example*

A large aluminum works ran five shifts all year long, with the shifts divided into teams operating one furnace each. About a year ago, a new pay system was introduced whereby the teams were continuously measured on their output, energy use, defect rate, and scrap metal use. Pay for the entire team was linked to performance along these dimensions.

The pay system was well liked, but there had been complaints that the previous shift filled the furnace with scrap metal. This made the first shift look good in terms of scrap metal use, but lowered the output levels for the following team. There were also complaints about poor cleaning, required maintenance not performed, vehicles parked haphazardly, and so on.

Believing that the pay system, although having raised productivity by close to 2 percent, was the cause of the trouble, the system was terminated in the early spring. After a few weeks of operation under the old system, there were more complaints than ever about sloppiness when leaving a shift.

Baffled, management ran a series of tests to pinpoint the reasons for this. They designed a number of scatter charts that linked the number of complaints with various causes.

One of the last charts revealed the culprit: As the scatter chart shows (Figure 16), there was a clear correlation between the number of complaints and the weather. It seemed that shift teams wanted to get off and into the good weather as soon as possible and were not properly cleaning up after themselves.

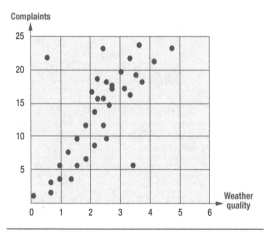

**Figure 16** Scatter chart example.

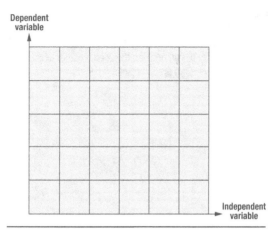

**Figure 17**   Scatter chart template.

*Histogram*

A histogram, also called a bar chart, is used to display the distribution and variation of a data set. The data can be measures of time, blood pressure, age, attitudes, and so on. The histogram's strength is in presenting data in a way that makes it easy to see relationships, which makes it useful at almost any stage of the RCA process. In root cause analysis, you can use a histogram to:

- Display data in a manner that makes it easier to determine which causes dominate

- Illustrate the distribution of occurrences of events and their causes and consequences

- Determine effects of implemented solutions

To create a histogram, follow these steps:

1. If the collected data has not been divided into categories, split them into a suitable number of categories (for example, periods during the day, age groups, types of causes).

2. Create a bar chart with space on the horizontal axis for the number of data categories; the vertical axis should accommodate the highest data point.

3. Create bars for each data point where height corresponds to the registered data.

4. Review the resulting histogram to look for patterns. If you end up with a chart with few bars, all bars nearly equal, or a comb-like pattern, reassess the number of categories and the division of data into the categories.

*Histogram example*

A small-town newspaper used teenagers to deliver the paper to subscribers. Frequent complaints about late deliveries suddenly started to occur from the area of one particular paper route. When staff confronted the paperboy with complaints, he was surprised, had no good explanation for the delays, but promised to keep up the standards.

After a brief period of significantly reduced complaints, they picked up again to the old level. The distribution manager asked a sample of subscribers on the route to make a note of every time the paper was delayed, and by how much. After four weeks of registration, the distribution manager analyzed the data (Figure 18). When confronted, the paperboy confessed that on Mondays, Fridays, and Saturdays his sister did the route for him. His sister was less familiar with the route and a slower rider, and this caused delays of an average of 20 minutes. Figure 19 is a histogram template.

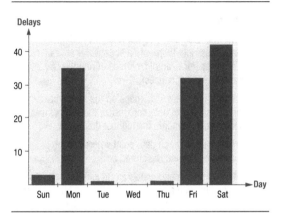

**Figure 18** Histogram example.

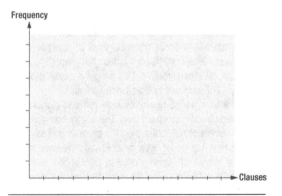

**Figure 19** Histogram template.

*Problem concentration diagram*

Where an event occurs may be important information. For example, in which part of the manufacturing facility do incidents occur? A problem concentration diagram is helpful in connecting events to physical locations and thus perhaps revealing patterns of occurrence. Proceed as follows:

1. Design the diagram by drawing a map of the building, area, or system.

2. Determine whether location-based event occurrence data already exist. If yes, skip to step 4.

3. If not, define what events are to be recorded and collect data linking events to locations.

4. Where more than one event is recorded, assign symbols to each.

5. Using collected data, plot events on the diagram.

6. Analyze the diagram to identify patterns of event occurrences.

*Problem concentration diagram example*

A large clothing store saw losses due to theft increasing steadily, despite alarms attached to about half of the garments displayed. After having caught a thief red-handed one day trying to put a sweater into a shopping bag from another store, one of the employees realized that one part of the store was vulnerable. It was hidden from easy view from the checkout counter and could not be observed using security cameras.

After collecting data about where garments were taken, the store constructed a problem concentration diagram. An employee drew a map of the store layout and located stolen garments in the map (Figure 20). It quickly became obvious that almost all items were stolen from areas with the same type of vulnerability. The alarm procedure was changed to include alarms on all items displayed in such areas. Although thieves still take garments from these areas by tearing off the alarms, the problem has been significantly reduced.

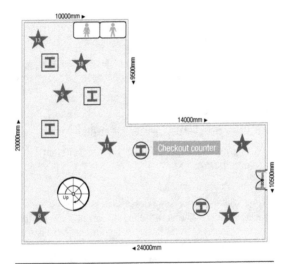

**Figure 20**    Problem concentration diagram example.

### 3d: Collate the Findings

Although it might seem trivial, this substep is important because it brings together "loose ends" from potentially many different types of analyses. Depending on how many different analyses you have performed in the previous substep, collating the findings can range from doing almost nothing to consolidating the findings from several exercises. Especially where different analyses indicate conflicting conclusions, you must make sure no wrong assumptions are made before proceeding. This can involve having to repeat analyses or collect more data. Conclude this substep by updating the cause-and-event tree from substep 3b, making sure all knowledge gleaned about the event is represented.

## Checklist for Step 3: Find the Root Cause

☐ The possible causes have been sorted into logical categories.

☐ Similar, overlapping, or related possible causes have been grouped together.

☐ A preliminary cause-and-event tree has been compiled based on the possible causes from the previous step.

☐ Relevant analyses have been performed to find root causes among possible causes.

☐ The results from different analyses aimed at finding root causes have been collated. The cause-and-event tree has been updated with the conclusions from the analyses.

☐ The assumed root causes have been critically
     discussed to ascertain whether they actually
     represent the true "root."

☐ The RCA team is content that the true root causes of
     the event have been identified.

## STEP 4: FIND SOLUTIONS

Having identified the actual root cause, this step in the RCA
process is about finding solutions that eliminate the root
cause. If you think you can relax the problem-solving effort
after having identified the root cause, think again! Your project
is not at its end until the root cause has been eliminated by
finding a solution that will prevent the event from reoccurring.

### Purpose of Step 4: Find Solutions

• The purpose of this step is thus to design workable
  solutions that eliminate the root cause.

**Success Factors in This Step**

• Keep up the momentum. Once the root cause has
  been identified, some people tend to think the process
  is finished. Focus on maintaining motivation and pace
  throughout the complete process of implementing and
  testing the effectiveness of the solution.

• Involve those with ideas about possible solutions.
  The RCA team may not have sufficient knowledge
  about all feasible solutions and should therefore
  involve a wide range of people who might have
  useful knowledge.

- Involve those who will be affected. People who will live with the solution must be satisfied with it and the process leading up to it. If not, they might passively or actively oppose it.

- Create ownership of the required changes. If the people involved feel that this is their solution, you have succeeded.

## Substeps in Step 4: Find Solutions

a. Explore the root cause; that is, review the bigger picture in which the root cause plays a part and consider the stakeholders who relate to the root cause.

b. Identify one or more solutions by using various creative techniques in a structured way.

c. Specify/describe the solutions, making sure that the conclusions from this step are well documented.

Step 4 has only three substeps, which makes finding solutions seemingly the shortest of the steps in the RCA process, but do not let this fool you; this step is often harder than you might think. Most people involved in RCA processes seem to relax once the root cause has been found. Both emotionally and cognitively, they view finding the root cause as the main challenge. You must counteract this attitude or you risk the entire process running out of steam before you find feasible solutions.

## 4a: Explore the Root Cause

Although the last step concluded with an identified root cause, it is often necessary to understand it better to find solutions. This means looking at the bigger picture and the stakeholders that surround the root cause. Relevant techniques for doing so have already been described:

- Flowchart
- Interview
- Survey

We do not present these again in this chapter, even though the purpose of using them is slightly different here. We have, however, outlined a possible way to combine these techniques to explore the root cause:

1. Develop flowcharts for the processes that include or relate closely to the root cause.

2. Develop a stakeholder map to place the root cause inside a larger picture.

3. Use interviews in order to understand the interrelationships between the processes, stakeholders, and the root cause.

4. If applicable, conduct surveys in order to obtain even more information about the root cause and its occurrence pattern. In this exercise, make sure you include input from the various stakeholders of the root cause.

## 4b: Identify Solutions

This substep, just like substep 3c of analyzing possible causes, is hard to describe as one streamlined sequence of tasks. Extreme situations are common. Sometimes a solution is readily available; other times, it requires weeks of study.

We will present some techniques we know can help, but we emphasize that their use can vary greatly; sometimes it suffices to use one of them, in other cases you might have to combine several:

- Brainstorming, now applied to finding solutions rather than causes
- Benchmarking
- Applying the "Why Not" principles

*What is a solution?*

To many the word "solution" sounds like a tangible thing, often in the form of physical equipment or features. In this context, a solution can take on many different shapes, for example:

- A new routine or process
- Computer software
- Computer hardware
- A "gadget"
- New competence and skill
- A quality system
- A foolproofing device
- Increased staff accountability

Keep this in mind when searching for a solution; it might make finding one easier.

*Benchmarking*

As with other tools explained in this book (for example, brainstorming), benchmarking is probably something many readers have heard about. In some cases, benchmarking might even carry a negative reputation as a numbers game where companies or departments or employees are "put in the stocks" for below-average performance.

For us, the essence of benchmarking is to learn from others. In root cause analysis, so-called process benchmarking can be used to learn from comparable processes, within the organization or externally. Very often, you will find that other units inside your own organization, or in other organizations, have encountered the same problem you have, and that best practice solutions have been developed to prevent it. Instead of (re)inventing the same or new solutions, benefit from this and learn how to apply the best practice solution.

To use process benchmarking to find solutions, make use of these general activities:

1. Brainstorm possible benchmarking partners (we recommend using at least two), that is, internal units or external organizations that could teach you about possible solutions.

2. Obtain agreement from the identified partners about their participation in the benchmarking study; be prepared to offer them information in return.

3. Study the benchmarking partners by interviewing them, visiting them, or reading their published best practices.

4. Compare the findings from the partners (if more than one was studied).

5. Discuss whether the resulting solutions could be applied to eliminate your root cause.

*Example of Benchmarking*

A telecommunications company experienced several problems with their efforts for measurement of customer satisfaction. To find the best possible benchmarking partners, the company defined a list of criteria for suitable organizations and their customer service:

- Profitability the last five years, as this was seen as an expression of the degree of customer satisfaction

- Multiple market segments, to find partners operating in a situation similar to the company

- A service industry niche, as customer satisfaction measurement in the service industries was considered quite different from what was being done in manufacturing organizations

- Long-term customer relationships, as opposed to companies dependent on one-time sales, as long-term relationship would enable a continuous measurement of customer satisfaction

- A technology-driven field, to resemble the company as much as possible

- Changing regulatory conditions, because such conditions impact customer satisfaction and the measurement of it

- Leadership in customer satisfaction

- Active use of feedback from the customers for process improvement

- A quantitative and systematic approach to measuring customer satisfaction

- Use of several different instruments for customer satisfaction data collection

In the end, eight organizations were agreed on as objects for further study. The selected partners belonged to industries such as banking, telecommunication, insurance, and public relations. To collect information, a questionnaire was produced consisting of two parts. The quantitative part focused on figures for the number of employees who performed various tasks, associated costs, the number of customers followed up, response rates for measurements, and so on. The qualitative part was far more extensive and focused on how customer satisfaction measurement was performed and applied and by whom, specific performance measures used, and so on.

Analysis of the data led to a list of recommendations regarding issues that should be changed:

- Establish an organization that could handle all customer responses and use it to improve products and services

- Terminate the generation of customer satisfaction data on the level below managers, to avoid employee fear of repercussions

- Stop basing the payment for lower-level managers on customer satisfaction data

- Develop internal process indicators linked to customer requirements

- Expand the scope of customer satisfaction measurement, but reduce the measurement frequency to once every three months

- Survey both customers who had recently been surveyed and those who had not, while also trying to reach the customers who rarely gave feedback

- Use customer satisfaction data at a strategic level

- Eliminate frustration at lower levels in the organization that results from being held responsible for measures an individual could only partially control or impact

*The "Why Not" Principles*

Nalebuff and Ayres[1] have created four approaches that act as catalysts for developing solutions to problems. Each is represented by a question (Table 7):

- *What would Croesus do?* (Croesus was the extremely wealthy king of Lydia ca. 560–546.) What solutions could you come up with if you were unconstrained by financial considerations? Thinking how an unconstrained person would solve the problem allows you to be a bit bolder and more outrageous than you might otherwise be. Typically, solutions prompted by this question will not be feasible in real life, but might represent a core idea that can be expanded upon.

- *Why don't you feel my pain?* This slightly cryptic question recognizes that individual and corporate actions have consequences to others that are not priced in the market (economists call these negative externalities). Looking for inefficient behavior by buyers or sellers is a systematic way both to identify problems and to solve them. We can identify problems by looking for behaviors that create an external harm that is greater than the internal benefit.

- *Where else would it work?* This approach builds on the fact that often a great solution exists for a different problem, one similar enough to your problem that the solution can be an inspiration. This normally requires some translation to fit the context and institutions of the new setting.

- *Would flipping it work?* There are symmetries all around us and sometimes flipping things around provides a powerful new solution. This is done by breaking down the existing practice into its component parts and writing a description in simple, declarative sentences. Then imagine what it would mean to turn around each or several of the components, flipping nouns, verbs, adjectives/ adverbs, and sometimes two words at once.

*Example of the "Why Not" Principles*

A rather large grocery store received a number of complaints during recent months due to goods that carried expiration dates too close to customer purchase dates. Customers were annoyed when they had to throw away food. The grocery store tried to use a few of the "Why Not" principles to solve the issue (Table 6). The blank template is shown in Table 7.

| What would an "unconstrained" person do? | • Automate the grocery store shelves, thus always being able to remove groceries that are soon expiring<br>• Give items that are about to expire to food pantries and other charitable organizations<br>• Differentiate prices based on expiration dates |
|---|---|
| Why don't you feel my pain? | • Compensate customers who have bought items that are about to expire |
| Where else would it work? | • Consider how restaurants, cafeterias, and other parts of the food industry address similar problems |
| Would flipping it work? | • Mark down prices on food about to expire; customers save on food that would otherwise be wasted<br>• Compensate customers for the full price plus a premium for goods that expire within 24 hours |

**Table 6** "Why Not" principles example.

| What would an "unconstrained" person do? | |
|---|---|
| Why don't you feel my pain? | |
| Where else would it work? | |
| Would flipping it work? | |

**Table 7**    "Why Not" principles template.

### 4c: Specify/Describe the Solutions

There are several reasons why this substep, which you might think superfluous, is important:

- As you will see in the next chapter, you might decide to let someone other than the RCA team be in charge of implementing the solution. In this case, a thorough description of the solution is absolutely essential, even if there will be close interaction between the implementation team and the original RCA team.

- At this stage, you might have a good idea about the solution, but describing it in detail requires that you clarify any loose ends.

- Various stakeholders of the RCA process expect to be informed about the outcome of the analysis.

- Information is required by ISO 9001 or other systems, if solutions are implemented.

- Documentation is important for future processes and continuous improvement, and should include the history of the project.

## Checklist for Step 4: Find Solution(s)

☐ The RCA team has switched to a creative mind-set.

☐ The root cause has been explored using various tools; the team understands the process in which the root cause appeared and recognizes all stakeholders.

☐ Feasible solutions have been identified through the use of suitable creative techniques.

☐ The developed solutions have been specified and described in sufficient detail such that a new team could potentially proceed with implementation.

*Step 4 End Note*

1. Nalebuff, Barry & Ayres, Ian (2006) *Why Not? How to use everyday ingenuity to solve problems big and small,* Harvard Business Press.

## STEP 5: TAKE ACTION

This chapter provides details about the second-to-last step of the root cause analysis process: implementing the solution designed to eliminate the root cause.

### Purpose of Step 5: Take Action

The first four steps of the RCA process are very much an exercise within the RCA team. Step 5 is a more open effort directed at driving change in the organization. This requires reliance on other people and their acceptance of the solution. The ultimate purpose of this step is to implement a solution that eliminates the root cause and ensures that the event under analysis does not reoccur.

#### Success Factors for This Step

- Successfully involving stakeholders, gatekeepers, and those affected by the proposed changes and helping them recognize that people must adopt new practices, investments might be required in equipment or software, training might be required, and so on. Nothing happens unless people understand why things must change.

- Having patience to accept that extensive change in work practice does not happen overnight.

## Substeps in Step 5: Take Action

We have split this step into five substeps. This might seem like a lot for a step with such a clear purpose, but most of these are rather limited in the amount of effort and time needed:

   a. Analyze the implementation setting, that is, the climate for change.

   b. Decide how to organize the implementation effort.

   c. Develop an implementation plan covering activities, responsibilities, deadlines, and so on, and have it accepted.

   d. Communicate with and create ownership of the changes by those affected by them and by those who will implement them.

   e. Implement the solution.

### 5a: Analyze the Implementation Setting

In our experience, most RCA teams are highly focused when completing Step 4 (describe in detail the solution required to eliminate the root cause). Fewer of them adequately consider how various employees and other stakeholders will receive the solution. The purpose of this substep is therefore to take a "virtual step back" and view the solution in light of this.

Our strongest recommendation is that the team set aside time for such discussion. A technique called force field analysis is useful to guide an assessment of the implementation setting.

*Effectiveness of Change*

Stakeholder acceptance of the proposed solution is essential. The change process can be viewed as a formula, involving three elements:

- E = the effectiveness of the change process
- Q = the quality of the change approach
- A = the acceptance of the change among those involved

The formula is E = Q x A.

Studies have shown that all successful change processes have high values for both Q and A. The same studies show that most failed change projects also have high Q. This means that a "technically sound" solution is in itself no guarantee that the change will be successful.

*Impact effort matrix*

An impact effort matrix (Figure 21) is a tool for deciding which of possibly many suggested solutions should be implemented. It illustrates which solutions seem easiest to achieve and which provide the most effects. The steps in constructing an impact effort matrix are:

1. Retrieve suggested solutions from previous discussions.

2. Construct an empty diagram with effort required to implement the solution on the horizontal axis and impact of the solution on the vertical axis, and divide it into four quadrants.

3. Assess effort and impact and place each solution in the diagram according to these assessments. Use symbols, colors, or labels to identify each possible cause.

4. Solutions falling into the upper left-hand quadrant will yield the best return on investments and should be considered first.

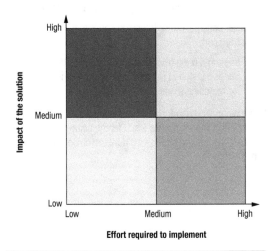

**Figure 21**    Impact effort matrix template.

*Impact effort matrix example*

A hospital saw a number of cases of patient identification errors. The events were investigated, and several proposed solutions surfaced:

- Create a policy for matching patient ID to chart ID at every point in transfer.

- Train every employee on patient identification policy.

- Bar code patient ID bands and charts; match before transport.

- Implant a radio frequency identification (RFID) tag under the skin of patients and on each medical chart for matching at each transfer point.

These were assessed for effort and impact, and plotted in an impact effort matrix (Figure 22). Clearly, using bar codes on patient ID bands and charts would be the obvious first option to proceed with.

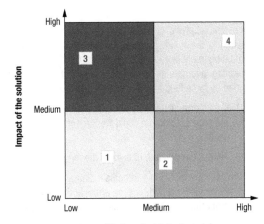

**Effort required to implement**

| Item # | Description |
|--------|-------------|
| 1 | Create a policy for matching patient ID to chart ID at every point of transfer. |
| 2 | Train every employee on patient identification policy. |
| 3 | Bar code patient ID bands and charts; match before transport. |
| 4 | Implant a radio frequency identification (RFID) tag under the skin of patients and on each medical chart for matching at each transfer point. |

**Figure 22**   Impact effort matrix example.

*Force field analysis*

Force field analysis is based on the assumption that any situation is the result of forces for and against the current state being in equilibrium. Countering opposing forces and/or increasing favorable forces will help induce change, and this is aided by force field analysis through the following steps:

1. Brainstorm all possible forces inside and outside the organization that could work for or against the solution.

2. Assess the strength of each of the forces.

3. Place the forces in a force field diagram (Figure 24), with the length of each arrow in the diagram proportional to the strength of the force it represents.

4. For each force, but especially the stronger ones, discuss how to increase the forces for change and reduce those against it.

*Force field analysis example*

During a reorganization debate in the local branch of a major volunteer organization, the issue of a common economy came up. Today a local branch of the organization has four departments, each with its own budget. Some argued that it should remain that way, while others argued that resources would be utilized better if they all shared joint accounts.

As the temperature of the debate rose, it was suggested that a force field analysis be used as a neutral tool to sketch the arguments for and against such a change (Figure 23). Although the arguments for change were important, it was decided not to change the current state, as the forces against dominated.

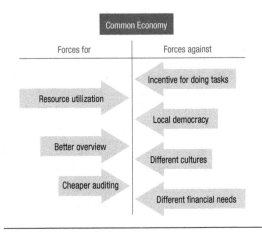

**Figure 23** Force field analysis example.

**Figure 24** Force field analysis template.

## 5b: Decide on the Implementation Organization

So far, the RCA team has conducted its work somewhat outside the regular running of the organization. From now on, the effort will "interfere" much more with ordinary operations. The RCA team could lead the implementation of the solution, but it's not necessary. The advantages and disadvantages of three alternatives are discussed here:

- The original RCA team takes charge of the implementation. The team knows the project, understands the proposed solution, and may be the unit best equipped to take care of implementation.

- A dedicated implementation team is developed, including members from the original RCA team and others necessary to ensure sufficient line management authority. This approach combines detailed insight into the root cause analysis with more formal authority and can be a sensible option in some cases.

- The unit organization, where actual organizational authority sits, drives change. In many cases this is the only real alternative. Trying to make change in work practice without support from affected unit managers can be a futile exercise. In fact, several generic improvement processes draw a line between analysis and solution development (improvement team responsibility) and change implementation (line management responsibility).

Although all three alternatives are options, the third option is the most frequently recommended.

## 5c: Create an Accepted Implementation Plan

This substep is partly related to the issue already discussed concerning having a project plan for the RCA itself. However, since the implementation of the solution often can have more far-reaching implications (investments, training, reorganization, and so on), the implementation plan is even more important. A typical implementation plan should cover:

- Implementation tasks and activities, with responsibilities and deadlines

- Resources to participate in and aid the implementation

- Cost estimate/budget, especially where implementation requires investments or other resource-intensive tasks

- Training needs

- Envisioned effects of the solution when implemented

We recommend using a Gantt chart to represent the tasks and project schedule for an implementation plan. However, the implementation plan is more than just the schedule; in fact, it often takes on the form of a project proposal. In essence, the RCA team is asking management at some level of the organization to sanction an investment of time, resources, and money. It might be useful to seek assistance from the finance department in calculating the return on such investments (ROI). An example of the structure of such a project proposal is illustrated in Figure 25.

| Section # | Content |
|-----------|---------|
| 1 | Background about the RCA project, the team, duration, mandate |
| 2 | The event investigated and the identified root cause |
| 3 | The solution designed |
| 4 | Implementation plan Gantt chart |
| 5 | Organization of the implementation and resources to be involved |
| 6 | Project costs, divided into categories such as investments, hours of employees, external services, and so on |
| 7 | Training needs to implement new practices |
| 8 | Estimated effects of the solution |
| 9 | If possible to develop, a cost/benefit analysis of the project |

*Total length not more than 8–10 pages.*

**Figure 25**   Implementation plan/project proposal example.

Unlike previous steps of the RCA process, at this point the RCA team must halt and wait for external confirmation before moving on. The project proposal must be presented to and sanctioned by management, a phase of the work that is frustrating to some teams. Satisfied that the root cause has been uncovered and a good solution designed to remove it, teams can become impatient as they wait for leadership approval.

One way to counter this situation is to request a meeting to present findings and the implementation plan instead of just submitting a written proposal. If you are lucky, approval can be obtained in that meeting. There is at least a chance to clarify issues not fully understood or to elaborate where those making the decision need more information.

## 5d: Communicate and Create Ownership

These are two distinctly different tasks.

Communication is about informing those affected by the implementation of proposed changes (employees involved in the actual process and also those interacting with it). Most of these people will be aware that there was an event and that an RCA team was established. Now is the time to follow up and present pertinent information about the status of the work and the solution.

Creating ownership requires targeting those who will need to change work practices as a consequence of the solution. Unidirectional communication flow is not enough; these people deserve an opportunity to be heard and to take part in discussions about the solution and its implementation.

*Overcoming resistance to change*

It's human nature to resist change and cling to the familiar. People in organizations are often reluctant to make the changes needed to implement solutions. Studies have found six common layers of resistance to change:

- Disagreement about the existence of a problem

- Belief that the problem is outside anyone's control

- Disagreement about whether the suggested solution can solve the problem

- Disagreement about whether suggested solutions will cause negative effects

- Creation of barriers against implementation

- Creation of doubt about whether others will cooperate in the solution

One way to overcome this resistance is to help people see that change is necessary and a smart thing to do; this is usually achieved through discussion and the influencing forces identified in the force field and barrier analyses. Another way is to employ "change agents," people who have standing and (often informal) authority in circles where resistance is expected. Having these people on board and agreeing to the proposed change will help you influence skeptics indirectly. Even better is including potential change agents in the RCA team from the outset.

## 5e: Implement the Solution

This is not a clean-cut task to be performed once and then ignored. This step can sometimes be a lengthy one, perhaps lasting several months. Some key activities are:

- Execute the implementation plan and follow up on its progress.

- In cases of deviation or delay, identify reasons and take action to remedy the problem.

- Be a positive force in driving the change required to implement the solution.

In essence, this is core project management. Please note that some of the discussion about Step 5 might be overkill; if the solution is that two nurses, not just one, be present when lifting a patient, chances are you will need neither force field analysis nor a Gantt chart.

## Checklist for Step 5: Take Action

☐ The setting for the coming implementation of a solution has been discussed.

☐ Forces for change and forces/barriers against it have been analyzed using either force field analysis or barrier analysis.

☐ Alternative ways to organize the implementation have been discussed and the best approach decided on.

☐ An implementation plan has been created and, if necessary, a more comprehensive change project proposal.

☐ The implementation plan/change project proposal has been sanctioned by those with authority.

☐ Those affected by the implementation have been properly informed about coming change.

☐ Efforts have been made to create ownership of the solution and required changes in those most influenced by it.

☐ The solution has been implemented as planned.

## STEP 6: MEASURE AND ASSESS

Here, finally, you will find an outline of the last step of the root cause analysis process, where you assess whether the solution actually eliminated the original problem.

### Purpose of Step 6: Measure and Assess

In some cases you will find that the proposed solution was not actually implemented or that the solution simply did not work. This makes this final step a necessity in order to properly complete the project and ensure that the action actually eliminated the problem that triggered the RCA process.

#### Success Factors for This Step

- Assessing the solution with a critical eye in order to be objective and unbiased

- Not being so keen to close the project that you might overlook important factors

### Substeps in Step 6: Measure and Assess

This final step of the RCA process contains five possible substeps. Substeps a and b are optional, depending on the event under investigation and the solution found. In many cases, the circumstances are so simple that there is no need for them. In others, however, we really do recommend undertaking them.

The substeps are:

a. Conduct a pilot study of the new solution.

b. Undertake measurement of the situation after implementation of the solution.

c. Assess the effects of the solution and determine whether further effort is required.

d. Put in place safeguards that ensure the event will not reoccur.

e. Report the results of the analysis and close the project.

## 6a: Conduct a Pilot Study

In some cases it might be beneficial to conduct a pilot study before implementing the RCA solution full scale (for example, if the solution would cause widespread change in organizational culture or if the organization is quite large).

It can take some time before the effects of a proposed solution materialize. In such cases, a pilot study is highly useful and can entail:

- Conducting "dry runs" where the new solution is tested without affecting customers

- Running the new solution for some time under close scrutiny, both to detect problems and monitor the effects

- Modifying the new solution if problems are found, or even looping back to previous steps of the RCA process

## 6b: Undertake Measurements

This is another substep that is sometimes completely superfluous. However, in other cases you will need to study and measure the effects in terms of different aspects and along several dimensions:

- Determine whether the solution works under various conditions.

- Measure change in the volume of problems seen before and after the root cause analysis was initiated.

- Undertake a balanced measurement to summarize the effects of RCA implementation.

### 6c: Assess the Effects

This substep is a checkpoint that allows you to address the following questions:

- Has the solution been successfully implemented?

- Do we believe, or do we have documentation, that the solution has eliminated the root cause and will prevent the event from reoccurring?

- If not, is there a need to recycle the entire RCA process or simply develop a new solution?

Answering these questions right after the implementation of a solution can be difficult; sometimes new practices need time to settle. In this case, either allow some time to pass or discuss this issue with stakeholders.

### 6d: Put in Place Safeguards against Event Reoccurrence

A peculiar fact about change in organizations is that things tend to revert to the previous state unless a dedicated effort is made to counter this tendency.

An entire field is dedicated to the study of change management and the introduction of change in organizations. When you have mastered the basics of root cause analysis, it is a field worthy of further study.

It is also a fact that a chain of events can circumvent safeguards you implement, so you must be vigilant in monitoring the situation and be prepared to invent additional solutions and safeguards.

*Safeguards*

*Safeguards* is a very generic term; we really want to think of mechanisms that promote the continued upholding of new approaches and discourage reverting to the old. Such mechanisms can take on many different shapes and forms, for example:

- Physical boundaries where appropriate, such as devices that make it impossible to go underneath something during lifting operations

- Incentives that motivate people to uphold new practices: financial incentives in the form of bonuses, higher budgets, and so on, but also non-financial incentives such as more free time, better food in the cafeteria, or a new couch for the staff room

- Training in new ways of working, to cement change and help people understand the benefits

- Documentation and measurement of the effects of new approaches, to reinforce how the organization will benefit

- Best practice descriptions, such as manuals that can be shared throughout the organization to expose other units to new solutions and possibly influence their adoption elsewhere

- Discussions in a larger audience, where people involved in RCA projects and units where new practices have been implemented can discuss how things work, whether further changes are required, and so on

## 6e: Report and Close the RCA Project

In Step 1, the RCA team was given a mandate. At the end of the job, it is time to report on the work to stakeholders:

- The person or unit that gave the team its mandate

- Line management of any units/processes that were involved in the work and the solution

- Regulatory bodies requiring a formal report

- Other external stakeholders that would find it reasonable to receive an update about the event and its handling

- Other similar units, either internal or external to the organization, that might wish to apply the learning from your work

For some of these, specific reporting formats and communication modes are required; for others, a written final report from the RCA project is in order. Although writing such a report can seem tiresome, it makes sense to close a project properly. The report is also valuable for future RCA teams.

*The final report*

We hope we are not scaring you with this talk about a final report. Many are not comfortable writing reports, and we are certainly not encouraging a scientific paper. But we know of cases where excellent root cause analysis work had to be repeated. Because no report was written, the organization found itself investigating a similar event a couple of years later. Most lessons learned from the previous exercise were forgotten, and a new RCA team had to reinvent the entire process. A final report from an RCA project can be as simple as just a few pages, but it should include the following (many of these overlap the implementation plan/project proposal):

- An introduction describing the mandate, the team, and the time frame

- A short description of the event and its consequences

- A brief description of the analyses done, recounting the steps taken during the project

- The identified root cause and the solution designed to eliminate it

- A description of the implementation including the challenges, the approach to used to solve them, how long it took

- The effects of the implementation, most importantly whether the root cause was removed

- Lessons learned, both positive and negative; knowledge that future RCA teams could benefit from

## Checklist for Step 6: Measure and Assess

☐ The need for a pilot study has been discussed; if warranted, a pilot study has been performed.

☐ If required, the effects of the solution have been measured.

☐ The effects of the implemented solution have been assessed, especially the likelihood that the solution will prevent the event from reoccurring.

☐ Safeguards to prevent reoccurrence of the event have been devised and implemented.

☐ Appropriate stakeholders have been informed of the project and its results.

☐ A final report from the RCA project has been written and distributed to relevant recipients.

☐ The project has been formally closed and the RCA team disbanded.

# Section III

## RCA EXAMPLE

### Overview

Carry Me Home Shopping Bags (CMHSB) is a small group of five manufacturing units spread across the Benelux countries (Belgium, the Netherlands, and Luxembourg). The group is built around one main product: plastic shopping bags. These come in a variety of sizes and designs, making them suitable for anything from small specialized shops to large grocery store chains, which account for about 60% of the sales.

In this example, we will concentrate on one of the three manufacturing sites. It is a "lean" unit: one managing director, two "multi-purpose" managers, and three shifts of eight factory operators. Annual revenue is about $8.1 million and there is pressure on the units to stay competitive; the threat of being closed down and having production transferred to one of the other sites ever present.

Carry Me Home shopping bags are made through a four-step process:

1. Extrusion of plastic granulate into film

2. Printing of film with company logos, text, and so on

3. Conversion, i.e., cutting the film into bags and welding the bags' seams

4. Packaging into cardboard boxes ready for shipping

The entire process is heavily automated, requiring only eight operators to run a setup of three plastic film extruders and three integrated lines for printing, conversion, and packaging:

- The extruded film comes out of the machine as a "tube" that is laid flat and rolled up for intermediate storage.

- The roll is three times the width of a finished bag, approximately 59 inches (150 cm).

- The remaining three steps run in-line from film on roll all the way up to sets of 50 folded bags packed 20 in each cardboard box.

- At this line, the speed is about 110 yards per minute (150 meters per minute), leaving little room for error in equipment or material.

In general, production ran smoothly with a unit/hour count among the highest in the group. However, intermittently, the conversion line would stop due to problems with material flow through the machines. It was determined that root cause analysis was needed to solve the problem and the team launched a traditional RCA process.

## Define the Event

The film would break and jam and the welding became uneven. With each stop, cleaning up and restarting could take anywhere from two minutes to three hours, lowering productivity dramatically. Despite much effort to adjust various process parameters, the problem just would not go away.

For weeks, everything would run nicely. Then, out of nowhere, problems would occur until suddenly things started to work again. It came as a relief each time things unexpectedly worked again, but it left everyone none the wiser and just as vulnerable the next time.

## Find Causes

After adjusting every possible process parameter, changing parts, and making sure the raw material was stable, the entire staff was truly perplexed about the cause of the difficulties. This was one of the first companies in the country to start plastic manufacturing back in 1965 and there were few people around with more expertise, so there was really no-one to call for help.

At first unaware of more systematic problem solving techniques, operators and managers started brainstorming about possible causes that had been overlooked.

A type of flowchart, in the form of a layout diagram for the manufacturing process, was constructed and studied closely (Figure 26). One idea that quickly came to mind was compatibility problems between extruder and conversion line. Rolls from one extruder appeared to cause problems in one conversion line but not on the other two. It took the group two tries to find the true root cause of the problem.

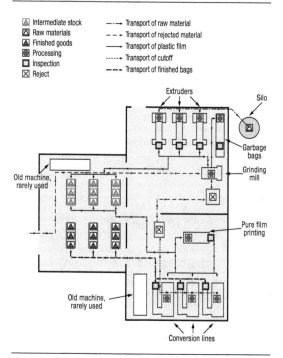

Legend:
- △ Intermediate stock
- △ Raw materials
- △ Finished goods
- ▣ Processing
- ☐ Inspection
- ☒ Reject

- –·–→ Transport of raw material
- – – → Transport of rejected material
- —— Transport of plastic film
- ······ Transport of cutoff
- — — Transport of finished bags

Extruders

Silo

Garbage bags

Grinding mill

Old machine, rarely used

Pure film printing

Old machine, rarely used

Conversion lines

**Figure 26**  Layout diagram.

## Attempt 1

### Find the Root Cause

To test the hypothesis, an effort was made to identify which extruder–conversion line combination was in use when problems occurred. This was achieved by marking each roll of plastic film with the extruder number and simply recording that number when a roll messed up the line. To everyone's surprise, there were no patterns to be seen. Apparently random combinations caused the problems.

Getting nowhere with their first approach, the company realized a more systematic line of attack had to be devised. It seemed clear that something with the film caused the conversion lines to stop and that more data were required. The company investigated several factors:

- Film thickness was specified at 35 μ (1 μ = 1 thousand of a millimeter); this was the main parameter that could vary.

- All rolls had the right total weight, but that did not preclude thickness variation across the roll.

- It was decided to collect thickness data, using sampling to minimize the number of tests to be taken and a check sheet to record the data.

- To capture all possible variation, samples were to be taken "longitudinally," i.e., from different places along the length of the film, and across the width of the film roll.

To identify possible causes, the sampling strategy called for samples to be taken from rolls that caused machines to stop. When such a "bad" roll came up, it was pulled out of production and replaced by a new one. Samples were taken from the rolls that were removed, 11 from the visible front upper side of the film "tube" and 11 from the invisible underside (Figure 27).

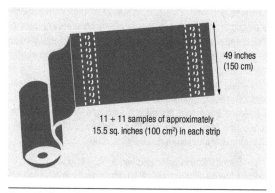

**Figure 27**    Data collection.

For each strip, the weight of the 22 samples was measured and, assuming a homogenous thickness across the circular sample, the thickness was calculated. The thickness data were then entered into a version of a check sheet, one for each roll of film. A simplified version of the check sheet is shown in Table 8.

Workers collected data this way for a period of two months and took samples from approximately 40 bad rolls stored during the last year, recording data from 590 strips (of 22 samples each) from 59 rolls. These data were analyzed, partly by calculating key statistical parameters (averages and standard deviation) and partly by using a histogram to portray thickness profiles across the width of the film.

The numbers in Table 9 are weight data for one 22-sample strip, in milligrams (mg).

**Extruder No.:** _____   **Date:** _____

| Strip # | Sample # | | | | | | | | | | | | | | | | | | | | | |
|---|---|---|---|---|---|---|---|---|---|---|---|---|---|---|---|---|---|---|---|---|---|---|
| | 1 | 2 | 3 | 4 | 5 | 6 | 7 | 8 | 9 | 10 | 11 | 12 | 13 | 14 | 15 | 16 | 17 | 18 | 19 | 20 | 21 | 22 |
| 1 | | | | | | | | | | | | | | | | | | | | | | |
| 2 | | | | | | | | | | | | | | | | | | | | | | |
| 3 | | | | | | | | | | | | | | | | | | | | | | |
| n | | | | | | | | | | | | | | | | | | | | | | |

**Table 8**   Simplified sampling check sheet.

**Extruder No.:** _____   **Date:** _____

| Strip # | Sample # | | | | | | | | | | | | | | | | | | | | | |
|---|---|---|---|---|---|---|---|---|---|---|---|---|---|---|---|---|---|---|---|---|---|---|
| | 1 | 2 | 3 | 4 | 5 | 6 | 7 | 8 | 9 | 10 | 11 | 12 | 13 | 14 | 15 | 16 | 17 | 18 | 19 | 20 | 21 | 22 |
| 1 | 38 | 38 | 35 | 38 | 39 | 35 | 35 | 46 | 43 | 40 | 40 | 38 | 37 | 37 | 34 | 35 | 38 | 37 | 37 | 34 | 37 | 37 |

**Table 9**   Weighted data for sampling check sheet.

For this strip, the key statistical parameters were:

- Average weight 37.7 mg.

- Average thickness 39.7 µ.

- Standard deviation 2.84 mg.

Comparing this with a strip from a "good" roll that ran without problems identified clear deviations:

- Average weight 37.2 mg.

- Average thickness 39.3 µ.

- Standard deviation 1.31 mg.

Looking further into the differences between good and bad rolls, workers could see that good rolls averaged 0.5 – 1.5 in standard deviation, with bad ones averaging above 2:

- Having identified a threshold value of 2 mg of standard deviation, workers implemented a procedure that called for a sample strip to be taken from every roll extruded

- Rolls of standard deviation higher than 2 were simply deemed waste, to be ground and reused as raw material

- This quickly eliminated most of the conversion line stops, but the root cause of the film thickness variation had not been found

- And although production ran better, the new procedure incurred extra costs in testing each roll and waste film that had to be reused

To further the analysis, this large thickness variation had to be better understood. It seemed clear that a histogram could be useful. By making bars for each of the 22 samples, workers generated one diagram per strip. Figure 28 is an example of such a histogram.

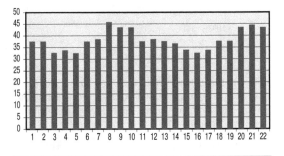

**Figure 28**    Film strip histogram.

The histograms identified an emerging pattern: two "peaks" of higher thickness and two "valleys" of lower thickness for each strip from bad rolls. The film comes off the roll as a flattened tube, and it was clear that two opposing areas of the tube were thicker and two opposing areas thinner, as shown on the next page.

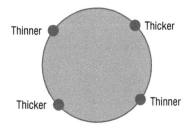

In the extrusion machine, what were there four of that could cause this variation pattern? An obvious element was the four-sided frame supporting the film tube and keeping it in a circular shape (Figure 29):

- The heated plastic is extruded through a tool (2).

- Air is blown from the inside of the extruded film tube, both to "inflate" it to its shape and to cool it.

- To stop the tube from expanding too much and to give it the circular shape, a supporting frame (3) is placed above the extrusion tool.

As the lower picture tries to show in more detail (Figure 29), the supporting frame consists of four arced steel rods with balls threaded onto them (much like an abacus):

- Together, these four rods form a closed circle.

- However, depending on the diameter of the tube being extruded, the diameter of the frame could be adjusted inward or outward.

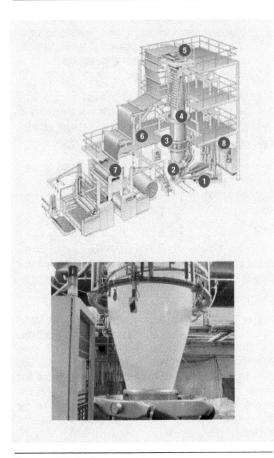

**Figure 29**   CMHSB extruder pictures.

- Looking at the geometry of the supporting frame, one could quite easily see that it formed a perfect circle at a medium diameter, but at lower and higher diameter had four irregularities matching the thickness variation patterns observed.

The theory explaining this was that in these four areas, the film was stretched a little more than elsewhere. Here, minor cooling variation would occur, introducing thickness variation.

### Find Solutions

To solve the problem, the supporting frame was adjusted to form a perfect circle at the diameter used most often, as much as 85% of the time. This should eliminate the root cause and enable the company to use all film rolls.

### Take Action

Confidence in the solution was great. Workers suspended the testing of all rolls even before production was resumed after the adjustment.

### Measure and Assess

Equally great was the surprise when exactly the same problems occurred shortly after! Having had time to digest the disappointment and assess the situation again, new measurements showed that the number of bad rolls had been slightly reduced. Adjusting the supporting frame had helped reduce the cooling variation somewhat. Still, this was obviously only an intermediate level cause and not the true root cause, proving again that root cause analysis often involves several iterations.

## Attempt 2

### Find the Root Cause

Having realized that the root cause hierarchy of this problem was more complex than first anticipated, workers next tried a five whys approach. A small team of people from different areas of the company convened to undertake the analysis. The data already collected were reviewed along with the analyses performed, and this led to the development of a Five Whys outline (Figure 30).

| Conversion line stops intermittently |
| :---: |

**Why?**   Some bad rolls of film appear

    **Why?**   Thickness variation across film width

        **Why?**   Uneven cooling of film during extrusion

            **Why?**   Cooling pipe moved out of position

                **Why?**   Operators turn pipe a little each time when cleaning off excess plastic at start-up

**Figure 30**   Conversion line Five Whys analysis.

During the session, it became clear that one of the operators knew about this problem and assumed everyone else did as well:

- Each time a new extrusion run was started, excess plastic had to be cleaned off around the extrusion tool.

- During the cleaning process, in a circular motion around the tool, the cooling pipe leading air upwards would typically be turned a quarter or a half turn, on fine threads.

- After a few such "treatments," the pipe would be lifted so much that the cooling airflow changed dramatically.

- The cooling air is channeled through a slit formed between two cones placed one inside the other, thus sending the air out in an upward direction (Figure 31).

- Screwing the inner part up too high caused air to flow directly outwards. When this happened, four ribs holding the pipe together at the slit became exposed and obstructed the airflow.

**Figure 31**   Geometry of extrusion tool.

It didn't take much analysis to see that these ribs matched the four points of thickness variation. The operator who was aware of the cooling pipe situation regularly adjusted it. None of the other operators ever did, which allowed it to reach this position and remain until the one aware operator again worked on the machine.

## Find Solutions

After this epiphany, the company felt confident that the true root cause had been found. Eliminating it would be a question of making sure the cooling pipe did not come so high that cooling was disrupted. There would probably be several ways to prevent this. The managing director, one of the other managers, and four operators were given a mandate to investigate. Several workable solutions emerged, with two that stood out as most promising:

- Implement a fixed routine to check the position of the cooling pipe every morning

- Install a locking pin on the cooling pipe, a pin that would have to be removed before any adjustment could be made to the pipe

A new routine would probably work quite well, but still allowed the potential for human error. If the locking pin could be designed, it would be a foolproof solution. While work started to design such a pin, the company also implemented a temporary inspection routine.

## Take Action

A small team of operators, with assistance from the extrusion machine supplier, set out to find a way to install a locking pin. It turned out to be quite easy, requiring only the drilling of a small hole through the threaded area of the pipe and the base. By making threads through the length of the hole, it was possible to mount a small lock screw to keep the pipe firmly in place.

The lock screw was easily removed to allow rotating the pipe for adjustment, cleaning, or servicing.

Actually implementing the new solution was thus purely a matter of making the required technical changes to the three extrusion machines. There was no need to create a change climate or assess forces opposing the change.

## Measure and Assess

Shortly after implementation of the lock screw on all three extruders, further film thickness measurements showed dramatic improvements with consistent thickness. The number of bad rolls has been reduced by 90% and the cost savings have been estimated at about $100,000 annually.

**Conclusion**

In these small pages we have described what root cause analysis is about, looked at an overall RCA process, examined the six steps of this process and applicable tools, and discussed an example illustrating the application of the RCA approach. We hope you take away from the text these main points:

- Root cause analysis is the best, perhaps only, approach to finding and implementing long-term solutions to problems or lower-than-expected performance, as it addresses not only intermediate causes but the primary trigger of the problem.

- Root cause analysis is a systematic process constructed to ensure that the outcome is an actual elimination of the root cause. Although the extent and level of detail included in each phase can vary from study to study, we do recommend following the logic of this process.

- Although root cause analysis is portrayed as a linear, straightforward process, be aware that real studies often require iterating a step, sometimes several iterations of several steps.

- The RCA process oscillates between analytic and creative modes of work, which can sometimes be frustrating or difficult for participants. However, the better participants are able to embrace these shifts in approach, the better will be the result.

- From reading this pocket guide, you might be tempted to think that conducting a root cause analysis is simply a matter of applying a number of tools in sequence. Yes, tools and analysis techniques are important in that they allow you to gather facts, look at those facts from different angles, promote creativity, and so on. But root cause analysis is as much about mindset. Successful root cause analysis depends on an organizational culture of continuous improvement that is always looking for problems or non-performance and truly desiring to solve these.

Where should you go from here to ensure that your organization becomes a professional executioner of root cause analysis? This of course depends very much on past experience in applying RCA and your current capabilities, but we offer some generalized advice:

- Provide training in root cause analysis. Although human beings are born curious and are normally geared toward improving non-satisfactory situations, very few people are born knowing how to most effectively improve. People with potential to become proficient in root cause analysis usually fulfill their potential if given a minimum of RCA training.

- Make RCA an every-day practice. We see many organizations where RCA is invoked only after serious incidents, perhaps even only when mandated by authorities or other bodies. RCA efforts triggered by external pressure or requirements are often conducted with limited motivation and inspiration and consequently suffer poor results. If RCA is a "once in a blue moon" exercise, the organization and its members never become proficient in the process. Conducting frequent RCAs—quick, simple exercises as well as more complex and time-consuming ones—ensures that a spirit of persistent vigilance and problem solving permeates the organization. Any organization will normally experience problems of such seriousness and complexity that a comprehensive RCA must be undertaken at frequent intervals. Don't wait for these problems to appear before conducting RCAs; apply the method to everyday issues that cause non-performance or annoyances.

- Include a broad selection of employees in the RCA process. Root cause analysis should not be reserved for an exclusive group of people. (In organizations where this is the case, this group typically comprises mainly engineers or other employees with higher education backgrounds in middle or higher management positions.) Successful RCA requires input from a variety of people and disciplines.

Involving a wide selection of employees means future RCA teams can draw members from a large pool of experienced participants.

This pocket guide should provide valuable support in both training efforts and ongoing RCA projects. Follow the steps of the RCA process, apply the tools and techniques described here, make use of the templates where relevant, and seek inspiration from the many examples provided. If you encounter situations where this pocket guide comes up short in explaining concepts or tools, look for other books and resources that offer more detailed insights.

We wish you success in your root cause analysis endeavors!

**Index**

Note: Page numbers in *italics* indicate figures or tables.

# NOTES

# NOTES

# NOTES

# NOTES

# NOTES

Printed in the USA
CPSIA information can be obtained
at www.ICGtesting.com
LVHW020912090124
768434LV00006B/637